José Lyavila
João Carvalho

Information gaps for the conservation of terrestrial biodiversity

José Lyavila
João Carvalho

Information gaps for the conservation of terrestrial biodiversity

The importance of identifying mammalian gaps in the northeastern region of southern Africa (Moz, Mw and Tz)

ScienciaScripts

Imprint

Any brand names and product names mentioned in this book are subject to trademark, brand or patent protection and are trademarks or registered trademarks of their respective holders. The use of brand names, product names, common names, trade names, product descriptions etc. even without a particular marking in this work is in no way to be construed to mean that such names may be regarded as unrestricted in respect of trademark and brand protection legislation and could thus be used by anyone.

Cover image: www.ingimage.com

This book is a translation from the original published under ISBN 978-620-6-76190-7.

Publisher:
Sciencia Scripts
is a trademark of
Dodo Books Indian Ocean Ltd. and OmniScriptum S.R.L publishing group

120 High Road, East Finchley, London, N2 9ED, United Kingdom
Str. Armeneasca 28/1, office 1, Chisinau MD-2012, Republic of Moldova, Europe
Printed at: see last page
ISBN: 978-620-8-10427-6

THANK YOU

Firstly, to God.

Special thanks to my supervisor, Professor João Carvalho, for his trust, guidance and encouragement in this research. I would like to thank him for making materials available, for his support during the development of this research and, above all, for the influence he has had on my scientific training, from the time I was taught to the time I was tutored.

To all the teachers who have passed through this master's programme in Applied Ecology at the Lúrio University in Aveiro, Portugal.

Many thanks to everyone who has been part of this journey to the end of this stage. I hope you will continue to take part in the new stages that are yet to come, you are all very important to my professional growth and development.

Thank you to the 2018 Ecology Master's class, just the best Master's class anyone could wish for. The friendship and support of each and every one of you has made all the difference in making it through these intense years and writing the final dissertation. Thank you to all the incredible people I've met during this time and become friends with. Thank you so much.

To my family... Thank you very much!

SUMMARY

Drawing up conservation plans requires identifying gaps in knowledge about the occurrence of species. This information forms the basis for selecting areas for future biological diversity surveys, as well as understanding the factors that explain the presence or absence of species in a given location. For this study, by integrating existing data from the Global Biodiversity Information Facility (GBIF) platform, we analysed the distribution of mammals in the Conservation Units of the north-eastern region of southern Africa (Mozambique, Malawi and Tanzania), as well as the variables that govern it. Initial data filtering led to the exclusion of some records where information on the date of recording, latitude and longitude was missing. This made it impossible to use the 12,708 occurrence data records initially available on the platform. Using the International Union for Conservation of Nature's Red List, it was possible to assess the status of 34 species. The criterion for selection was the number of occurrences of each species, with a minimum of 15 occurrences per species. 710 records of species occurrences were used to visually analyse their distribution and assess the factors that may influence it. Different parameters, Shannon Wiener (H'), Simpson (D'), Hill and Jacard's similarity indices were used to verify the differences in the composition of the mammal community in the countries studied. The results indicate that the study area showed different values for the indices evaluated, with Tanzania showing a value of 3.03 and 0.95 for the Shannon-Weaver (H') and Simpson's (C') diversity indices, respectively. The visual and descriptive analysis of the data revealed a sampling bias which made it impossible to draw firm conclusions about the effect of environmental variations on the distribution of species. However, there are species that exploit the areas covered by the protected areas, but also the surrounding areas. This aspect requires effective monitoring activities with a view to implementing specific measures for the conservation and management of the natural values surrounding the protected areas.

Keywords: Distribution, mammals, environmental variables, diversity indices, conservation.

1. INTRODUCTION

North-eastern Southern Africa covers three countries (Mozambique, Malawi and Tanzania) and two main river basins (Zambezi and Rovuma). It is a region with high biodiversity and very heterogeneous ecosystems. The aim of this work is to analyse the bias and gaps in information regarding the distribution of mammals in the study area in time and space, and to determine, if possible, the environmental factors (e.g. climate and other relevant characteristics of the region) that influence the distribution of the species studied.

Mammals have an astonishing diversity of species, forms, life histories and behaviours. They provide us with numerous direct benefits by being, for example, an important source of protein for many crops and rural populations and indirectly by playing a crucial role in maintaining balanced ecosystems as prey, predators, pollinators and seed dispersers (Karim & Ahsan, 2016).

The East and Southern Africa region is one of the most biodiverse areas on the planet, consisting of a series of diverse protected and conservation areas managed by a wide range of actors, from governments, non-governmental organisations (NGOs), local communities, private institutions and partnerships between these entities. The region has high levels of poverty and unemployment and, for this reason, governments tend to focus on socio-economic development. In particular, the concentration on the agricultural and mining sector, as well as major infrastructural developments, can result in reduced investment and funding for protected areas and conservation (IUCN, 2021).

Spatial patterns of specific richness, such as the latitudinal gradient of biodiversity, are an important biogeographical attribute that provides us with information on the functionality of ecosystems and allows us to outline conservation and management measures at different spatial and temporal scales (Menegotto & Rangel, 2018). Describing spatial patterns of richness requires precise knowledge of the distribution of the taxonomic group of interest.

Determining whether the absence of occurrence records for a species in a given region is a true absence or a sampling artefact is one of the main challenges facing researchers. By identifying spatial gaps in the distribution of species, it is possible to explore how and where absences are affecting the estimation of species richness. On the other hand, the absence of a species in an area where a high sampling effort has been employed suggests a true absence, which is a valuable result for studying ecological and conservation processes (Menegotto & Rangel, 2018). Few studies have explored the spatial distribution of species occurrences in this region and assessed the effectiveness of protected areas in effectively protecting the mammal diversity of north-eastern southern Africa (Mozambique, Malawi and Tanzania), Effective conservation planning relies on in-depth knowledge and acquisition of data on species occurrence and distribution. Primary data on the occurrence of species in scattered data sources can be a resource for increasing knowledge about the biodiversity of a country or region. Filling data gaps is crucial to obtaining new and extensive knowledge for biodiversity research and conservation. Regions neglected by research, which lack up-to-date information, often coincide with species-rich and developing nations (Neves et al., 2019).

The use of biodiversity mapping techniques enables integrated communication between the different players involved in the conservation of natural resources. In order to understand the patterns of spatial distribution and variation of fauna, it is necessary to know the pattern of spatial and structural alteration of habitats in the region and the factors that act on the composition and dynamics of plant communities. The relationship between vegetation patterns and soil and topography variables can help define sites with high diversity and a high percentage of biomass in the region in question. In this research, however, vegetation structure and diversity data were not used due to the lack of detailed information for the region of interest (Lecours, 2017).

The development of this study based on the GBIF, WDPA and WORLDCLIM repositories, which are open and free of charge, allows us to learn about the environmental variables that influence the known distribution of species and to try to understand the importance of protected areas for their conservation. Based on these two objectives, this study has produced knowledge about the gaps related to the distribution of mammals in the region. In this way, the results could help to conduct future studies focused on wildlife monitoring and subsidise the conservation of mammals in the region. public policies in the countries involved in the research (Mozambique, Malawi and Tanzania), with regard to decision-making aimed at monitoring and preserving these species. Species lists are the alpha-diversity descriptors of the taxonomic richness of a given country. Checklists at national level are important for conservation strategies at continental level, as the effective conservation of threatened species can be ensured by governmental organisations at national level more than any other, local or non-governmental agencies. The comparison of species lists between countries can make it possible to classify the various countries according to their conservation value in certain geographical contexts.(Amori et al., 2012) As is the case with this study carried out in the north-eastern region of southern Africa, it is important to identify information gaps on the organisms present for conservation.

The spatial coverage of mammal location points in GBIF does not reflect the known global patterns of mammal richness (Schipper et al., 2008), since the maximum density of location points in GBIF is found in North America and Europe, while the maximum richness of mammal species is found in the tropics (e.g. Mozambique, Malawi and Tanzania). This is strong evidence of a global spatial bias in data collection, which is to be expected in a meta-database like GBIF because there is no systematic plan and means of data acquisition (Boitani et al., 2011).

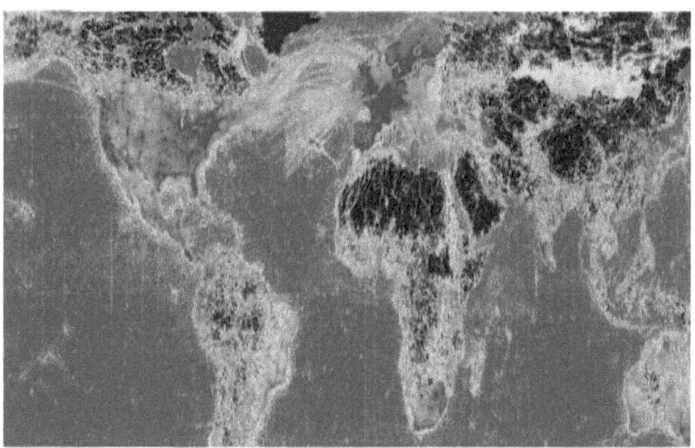

Figure 1 - The distribution of localisation points for all mammal species available on GBIF (GBIF, 2022).

The data currently available worldwide only partially cover the entire mammal class and are of five types: taxonomy (Schmid et al., 1993) phylogeny (Fritz et al., 2009) distribution (GBIF. 2011) & (IUCN Red List. 2011), life history and conservation status with supplementary information on threats. These datasets have been used in a variety of analyses, from large-scale biogeographical and macroecological studies (e.g. to local-scale conservation planning and attempts to predict the impacts of climate change on the risk of species extinction) (Boitani et al., 2011). Accurate maps of current species distributions are an essential component for assessing the conservation status of species (Freitag & Van Jaarsveld, 1998) & (Schipper et al., 2008) and for identifying conservation priorities (Eklund et al.,

2011) & (Wilson et al., 2011). Spatial conservation prioritisation analyses in particular, including gap analysis and the systematic identification of conservation sites, are sensitive to errors of commission (false presence on the map) and omission (false absence on the map) (Rondinini et al., 2006), 2006), with the first type of error being particularly problematic as it induces an underestimation of the number of gap species (species not well protected by existing protected areas) (Catullo et al., 2008) and the additional amount of area that needs to be protected to fill the gaps (Rondinini et al., 2005). Although human well-being essentially depends on the preservation of biodiversity to ensure the functioning of ecosystems (Naeem et al., 2016), current conservation efforts worldwide are failing to halt the decline of biodiversity, which is occurring at an unprecedented rate and will continue to do so due to current and new threats (Chen & Peng, 2017).

This situation has vast negative effects on the economy and society due to losses of ecosystem services and clearly demonstrates that the fundamental role of biodiversity for human well-being is not easily recognised by politicians who want to achieve these global goals. Despite recent efforts to uncover the synergies and delays between the goals and targets of these conventions, there remains a high risk that nations will choose some goals that suit their priorities and fail to achieve those that are more difficult to realise (Frank & Schäffler, 2019). While new indicators are constantly being developed and others improved or updated in terms of their availability to more countries (Frank & Schäffler, 2019), key knowledge gaps remain, especially with regard to data from developing countries. Identifying key knowledge gaps is important for assessing threats to biodiversity (Joppa et al., 2016) as well as for improving monitoring, management and investment strategies. Considering the socio-economic interest of the subject and the ecological importance of the taxonomic group in question, this study aims to answer the following questions:

• Which study area has the greatest gaps in information regarding the distribution of the taxonomic group identified in relation to conservation units?

• What factors condition the distribution of wealth in the region under study?

1.1. Problematisation and justification of the study

Due to a number of constraints, the availability of biodiversity-related information varies considerably in space, time, taxa and types of data, thus causing gaps in knowledge. Despite researchers' growing awareness of this issue, it is still unclear how - and if - scientific endeavours have contributed to filling these information gaps. Species distribution patterns are essential for understanding the possible ecological impacts produced by various human activities aimed at exploiting the environment. The loss of species in ecosystems can be caused by various anthropogenic activities such as burning, the opening up of agricultural fields, pastoralism and others that are already known and reported in various publications. Currently, the military conflicts taking place in the northern region of Mozambique also stand out, as do the neighbouring countries with which it shares a border (Mozambique, Malawi and Tanzania). There are various co-operations and agreements established in the management of shared natural resources in the north-eastern region of southern Africa.

African mammals are involved in conservation issues that are considered very problematic and challenging, particularly due to the rapid growth of the human population in this region associated with military conflicts that result in the plundering of natural resources (Laurance et al., 2014). The conservation of this taxonomic group is based on a combination of multiple threats and opportunities, requiring a good understanding of the biodiversity found in the region and the ecological conditions present, as well as how this group interacts with human populations in the context of the complex and dynamic ecological systems that characterise the study area (Huntley et al., 2019). The choice of these three countries, Mozambique, Malawi and Tanzania, is related to their high levels of natural resources, shared transboundary regions and threats and opportunities related to the conservation of biodiversity found in the aforementioned countries.

1.2. Objectives and hypotheses

Recently, advances in information technology, combined with the sharing of primary biodiversity data, are allowing unprecedented access to species occurrence data (Neves et al., 2018). By combining species presence data with digital mapping through a series of algorithms, the estimation of niches and distribution areas becomes feasible at resolutions one to three orders of magnitude higher than what was possible a few years ago (Peterson & Soberón, 2012).

The study of biodiversity distribution at different spatial and temporal scales has been one of the focuses of ecology and biogeography. Reliable descriptions of the distribution of species, as well as the identification of knowledge gaps, are fundamental to their conservation and management (Dormann, 2007). In this work, my general objective is to analyse the bias and information gaps relating to the distribution of mammals in the north-eastern region of southern Africa (Mozambique, Malawi and Tanzania). The specific objectives are:

- To analyse the spatial distribution of mammals in the north-eastern region of southern Africa (Mozambique, Malawi and Tanzania);
- Compare the current state of richness of the taxonomic group identified in the region using indices used in ecological studies and assess the impact of conservation units;
- Identify the variables that may influence the distribution of the taxonomic group identified.

2. MATERIALS AND METHODS

2.1. Area of study

The study area covers the Zambezi and Rovuma river basins and is closely associated with habitats characteristic of tropical and equatorial climates (Figure 1).

Mozambique is a country with a large geographical area (799,380 km^2) and is characterised by its enormous geological diversity, including fossils, inland lakes, swamps, marine and continental areas, various types of climate (tropical, subtropical, high altitude, semi-desert) and altitudes that reach 2436 metres at its highest point, Mount Binga, in the province of Manica. It is a southern African country located between the mouth of the Rovuma River and the Republic of South Africa, specifically between parallels 10°27' and 26°52' South latitude and meridians 30°12' and 40°51' East longitude. In the south-east of Africa there is a very small country (118,484 Km^2), Malawi. It borders Tanzania to the north, Mozambique to the south-east, east and south, and Zambia to the west. It lies between latitudes 9° and 18° S, and longitudes 32° and 36° E. Its relief varies between the plains of the Shire River, which originates in Lake Niassa and flows into the Zambezi River in Mozambican territory, and plateaus from the western border with Zambia to the vicinity of the western shore of Lake Niassa. In the southeast of the country, to the east of the Shire River valley, rises the Mulanje massif (part of the marginal Rift Valley chains) with Sapitwa Peak, which, at 3,022 metres, is the highest point in the country.

The climate is tropical in the central and northern regions, with an average annual temperature of 30 °C, and milder (temperate climate) in the south, under the influence of cold air currents (in winter) from the south of the African continent, with better defined seasons than the centre-north of the country (Happold, 1987).To the north of Mozambique is the United Republic of Tanzania which, with an area of 945,087 km², it is the 31st largest country in the world. It contains large, ecologically important wildlife parks, such as the famous Ngorongoro Crater in Serengueti National Park in the north. With an equatorial climate and an average annual temperature of 24°, this beautiful African country by the Indian Ocean is made up of a continental part and an island part with the islands of Pemba, Zanzibar and Mafia. Tanzania has the highest peak in Africa, Mount Kilimanjaro at 5891 metres, but also Lake Tanganyka and Lake Victoria, as well as the Serengueti Wildlife Park. With important natural ecosystems, it has an extremely varied biodiversity. Although it is a country world-famous for its safaris, it has habitats in two very important biodiversity hotspots: the Eastern Arc and the Albertine Rift Mountains surrounding Lake Tanganyika.

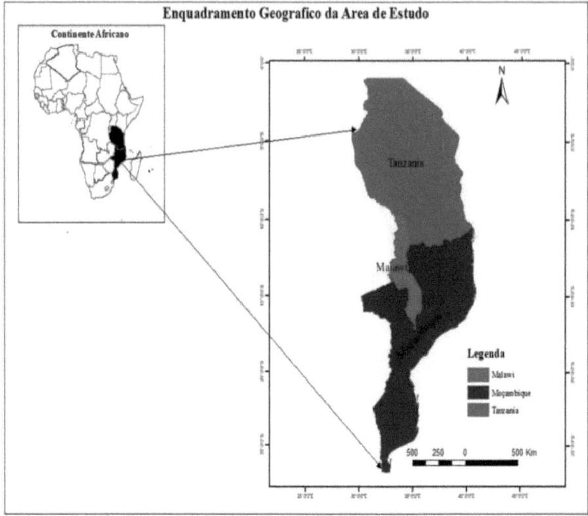

Figure 2 - Location of the study area.

3. DATA FROM PRESENCE

The study was conducted in the north-eastern region of southern Africa, using free biodiversity data downloaded from the GBIF (Global Biodiversity Information Facility) platform on 5 January 2021, with the reference https://doi.org/10.15468/dl.52f654, in three countries in the north-eastern region of southern Africa (Mozambique, Malawi and Tanzania). The choice of this international open-access platform, with its specific scientific structure, provides information to any person/researcher in the natural sciences, respecting the principles established within the organisation according to the agreements for the publisher and user of data. The 34 species in the taxonomic group identified in this study were chosen on the basis of the number of occurrence records for each species in the three sampling units (Mozambique, Malawi and Tanzania). The minimum number of occurrences was stratified and defined as 15 and the maximum number of occurrences as 30 for each species, this number representing the value found in Excel downloaded from GBIF (Table 2). The species were briefly analysed in the IUCN Red List to check the conservation status of each species.The 34 mammal species selected were classified according to the following categories: Not Evaluated (NA), Insufficient Information (II), Least Concern (PP), Near Threatened (QA), Vulnerable (VU), Critically Endangered (CR), (Appendix 1). Some mammal species have experienced long-term fluctuations. To understand the reasons for these changes, more systematic ecological, environmental and human data is needed to assess issues that may otherwise go unnoticed. One-off data collected in an unsystematic way can produce incorrect conclusions about the impact of vegetation succession on small mammal community structure, species interactions and resilience (Balčiauskas & Balčiauskienė, 2022).

3.1. Variables environmental

The distribution of organisms (plants and animals) has undergone changes, partially or totally, induced by changes in regional and global bioclimatic conditions. Current predictions from distribution models generally show changes in the level and variation of climate over the years and over the next few decades or generations. The WorldClim platform (http://worldclim.org) version 2.1 is a free climate database with high spatial resolution. The data provided was used to assess the factors influencing the distribution of the mammals studied. Nineteen bioclimatic variables are available in four spatial resolutions (expressed in minutes): 10 minutes, 5 minutes, 2.5 minutes and 30 seconds. To test the effect of the environment on the distribution of mammals in the study area, we used two environmental variables in the table below: average annual temperature and annual rainfall (BIO01 and BIO12).

Table 1 - Bioclimatic variables.

Order	Code	Description Variable
01	BIO01	Average annual temperature (°C * 10)
02	BIO02	Daily temperature oscillation (°C * 10)
03	BIO03	Isothermality (%)
04	BIO04	Thermal seasonality (Standard deviation * 100)
05	BIO05	Maximum temperature of the warmest month (°C * 10)
06	BIO06	Minimum temperature of the coldest month (°C * 10)
07	BIO07	Annual temperature oscillation (°C * 10)
08	BIO08	Average warm season temperature (°C * 10)
09	BIO09	Average dry season temperature (°C * 10)
10	BIO10	Average wet season temperature (°C * 10)
11	BIO11	Average cold season temperature (°C * 10)
12	BIO12	Annual rainfall (mm)
13	BIO13	Precipitation of the wettest month (mm)
14	BIO14	Precipitation of the driest month (mm)
15	BIO15	Rainfall seasonality (Coefficient of Variation)
16	BIO16	Wet season rainfall (mm)
17	BIO17	Dry season rainfall (mm)
18	BIO18	Warm season rainfall (mm)
19	BIO19	Cold season rainfall (mm)

Given today's rapid climate change, it is essential that we identify the climatic conditions tolerated by mammals in the study area, as well as which man-made landscapes limit the occurrence of species.

3.2. Analysing data

3.2.1. Indexes from diversity

ArcGIS 10.5 software was used to analyse the spatial distribution of mammals in the study area. The R software (R Core Team, 2022) was used to estimate richness at the sampling sites (Mozambique, Malawi and Tanzania) with the parameters illustrated below:

Simpson's index. This is an index of dominance and reflects the probability that two individuals chosen at random in the community belong to the same species. It ranges from 0 to 1 and the higher it is, the more likely it is that the individuals are of the same species, i.e. the greater the dominance and the lower the diversity. It is calculated as:

$$D = \frac{\sum_{i=1}^{s} n_1 (n_1 - 1)}{N(N - 1)}$$

Where:

D = Simpson's index

S = For i ranging from 1 to S (Wealth)

n = Total number of organisms of a species

N = Total number of organisms of all species

Shannon Index. It measures the degree of uncertainty in predicting which species an individual chosen at random from a sample of S species and N individuals will belong to. The lower the value of the Shannon index, the lower the degree of uncertainty and, therefore, the diversity of the sample is low. Diversity tends to be higher the higher the value of the index. It is calculated using the formula:

$$H' = -\sum_{i=1}^{s} p_i \log_b p_i$$

Where:

H' = Shannon index

"pi" = the frequency of each species

"S" = the number of species sampled for i ranging from 1 to S (Richness). "b" = the base of the logarithm (Uramoto et al., 2005)

The comparison was made between the countries involved in the study. The **Jaccard similarity index** (JSI) was used to assess the similarity between sampling environments, based on the number of common species (Ferreira et al., 2008). The index is defined by the formula:

$$Sj = \frac{a}{a+b+c}$$

Where:

Sj = Jaccard's coefficient;

a = number of species present only in sample a;

b = number of species present only in sample b;

c = number of species common to both samples.

In general, the values for species richness, Shannon Index (H') and Simpson Index differ basically in the weight given to rare species. We can generalise this idea using diversity profiles such as the Rényi or Hill series (Melo, 2008).

Where:

$$H_\alpha = \frac{\ln(p_1^\alpha + p_2^\alpha + p_3^\alpha + \cdots . + p_s^\alpha)}{1 - \alpha}$$

H_a = is the value of the diversity index for parameter α ($\alpha > 0$, $\alpha \neq 1$);

ln = Natural logarithm and

p_1, p_2, p_3 ... , p_n = are the proportions of individuals of species 1, 2, 3 ... S. E is given by,

$$N_q = \left(p_1^q + p_2^q + p_3^q \cdots + p_s^q\right)^{\frac{1}{1-q}}$$

Where:

N_q is the value of the diversity index for parameter q. When q = 0, N_O is equal to species richness (S).

4. RESULTS

4.1. Assessment of the richness of the taxonomic group mammals

The analyses carried out included 34 species of mammals (mainly micromammals), with a total of 710 occurrences in the north-eastern region of southern Africa (Mozambique, Malawi and Tanzania).

Table 2 - Number of records per species per country.

Species	Mozambique	Country Malawi	s Tanzania	Total
Aethomys kaiseri			17	17
Aethomys nyikae		16		16
Arvicanthis neumanni			24	24
Beamys hindei			23	23
Chaerephon pumilus	17	16		33
Colobus angolensis			19	19
Colobus guereza			20	20
Congosorex phillipsorum			24	24
Crocidura mdumai			17	17
Crocidura munissii			15	15
Crocidura olivieri		22		22
Cryptomys hottentotus		17		17
Elephantulus brachyrhynchus		15		15
Epomophorus crypturus	15			15
Grammomys cometes	16			16
Graphiurus microtis			15	15
Graphiurus murinus		30		30
Hipposideros vittatus			21	21
Miniopterus natalensis			21	21
Mops bakarii			21	21
Mus musculoides			25	25
Neoromicia nana			17	17
Nycteris thebaica	26		19	45
Otomys denti		15		15
Pelomys minor			25	25
Petrodromus tetradactylus			15	15
Rhinolophus clivosus		15		15
Rhinolophus deckenii			27	27
Rhinolophus hildebrandtii	19			19
Rhinolophus mossambicus	20			20
Rhinolophus swinnyi		24		24

Rhynchocyon petersi			21	21
Suncus megalura			17	17
Triaenops afer			24	24
Total	113	170	427	710

Table 3 - Diversity indices and profiles of the mammal taxonomic group.

Nature (species) Index de Jaccard/Similarity Biodiversity biodiversity

	fi	fr	Exhibition	Intensity	Simpson	Shannon	Malawi	Mozambique
Mz	113	15.92%	6	1066	0.88	1.77	0.07	
Mw	170	23.94%	9	2624	0.83	2.16		
Tz	427	60.14%	19	10415	0.95	3.03	0.00	0.04
Total	710	100%	34	14105	0.97	3.49		

Having carried out an exploratory analysis of the data by country (Mozambique, Malawi and Tanzania), the **Shannon** index indicates that the Tanzanian community is more diverse (3.03) than the Malawian community (2.16) and Mozambique (1.77), with the Malawian community (2.16) being more diverse than Mozambique (1.77). On the other hand, **Simpson**'s index indicates that the community in Mozambique is more concentrated (0.88) than in Malawi (0.83) and Tanzania (0.95), with Malawi being more concentrated than Tanzania (0.95).

Table 4 - Hill series.

q	0	0.25	0.5	1	2	4	8	16	32	64	Inf
Mw	9	8.93	8.86	8.71	8.41	7.83	7.00	6.35	5.99	5.82	5.67
Mz	6	5.97	5.95	5.90	5.79	5.56	5.18	4.79	4.56	4.45	4.35
Tz	19	20.92	20.83	20.67	20.37	19.82	18.99	18.03	17.17	16.52	15.81

4.2. Grouping of species by sampling community (level of similarity)

The similarity index, also known as the similarity coefficient, presents the level of similarity between communities in a simple and clear way. It is therefore a method used in ecological studies. In comparison, there is the dissimilarity index, which reflects the distance (not similarity) between communities. Both indices are related to beta diversity, measuring the extent to which species composition varies between communities. The results show that there is similarity between the units in the sample. This study analysed 34 mammal species in the north-eastern region of southern Africa, represented by three communities: Mozambique, Malawi and Tanzania. The Jaccard similarity index (SJ) was used to find the relationship between common species and the total number of mammal species found in the sampling region when analysing the number of species found in the region. compare the three communities. It can be seen that the communities in Malawi and Tanzania are the most similar (Figure 2).

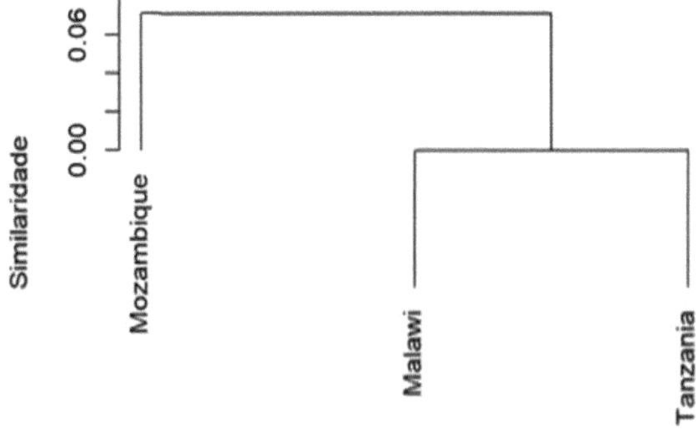

Figure 3 - Dendrogram based on Jaccard's similarity index between the communities (Mozambique, Malawi and Tanzania) in the mammal taxonomic group.

4.3. Influence of temperature on the distribution of mammals

4.3.1. Mozambique

Mozambique has an average annual maximum temperature of 27.1°C and a minimum of 12.4° C. The lowest average annual temperature was recorded in the north (Niassa), centre (Zambézia, Tete and Manica) and south (Maputo), while the highest average annual temperature was recorded in the centre and north along the Indian Ocean coast. The occurrence and distribution records show that the mammals studied occur in the temperature range between 23 and 24° C, as is the case with the species Grammomys cometes and Rhinolophus mossambicus; from 24 to 25° C we find the species Epomophorus crypturus and finally in the ranges between 25 and 27°C we find species such as Epomophorus crypturus, Rhinolophus mossambicus, Chaerephon pumilus and Rhinolophus hildebrandtii. The map below illustrates the variation and influence of temperature on the organisms studied.

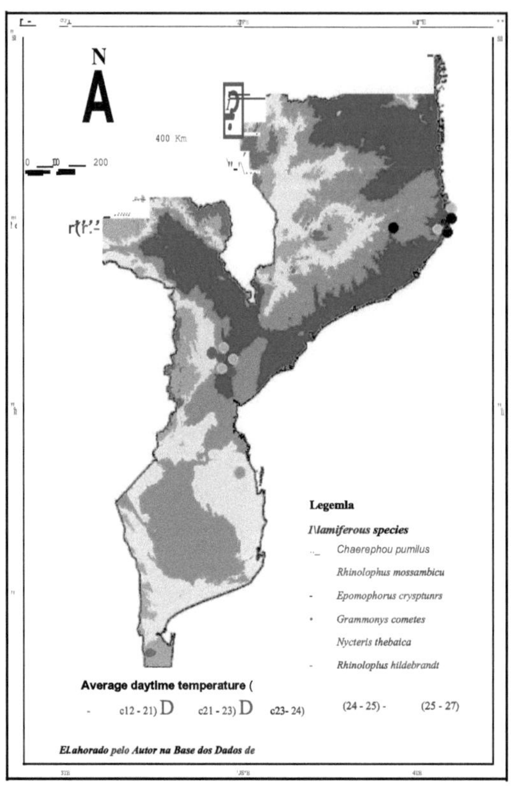

Figure 4 - Annual temperature variation generated in Arcgis 10.5 (Mozambique).

4.3.2. Malawi

Malawi has an average annual maximum temperature of 26.0°C and a minimum of 11.4° C. The lowest average annual temperature was recorded in the northern region in the provinces of Chitipa, Karonga, Rumphi and Nkhata Bay, with a range of 11-17° C, while the highest average annual temperature was recorded in the southern region along the coast of Lake Niassa with a range of 23-26° C. The records of occurrence and distribution of the organisms show that the mammals studied are present in all ranges of minimum and maximum temperatures in this sampling unit, varying between 11 and 26° C, with some specific cases of organisms that seem to tolerate a wider spectrum of weather conditions. The species Aethomys nyikae, Elephantulus brachyrhynchus, Chaerephon pumilus, Cryptomys hottentotus, Graphiurus murinus, Rhinolophus clivosus and Rhinolophus swinnyi are mostly concentrated in the north of Malawi in the provinces of Chitipa and Karonga, with temperatures ranging from 11 to 21° C and lastly, in the 23 to 26°C range, the species Aethomys nyikae was found in the southern part of the country in Mangochi province.

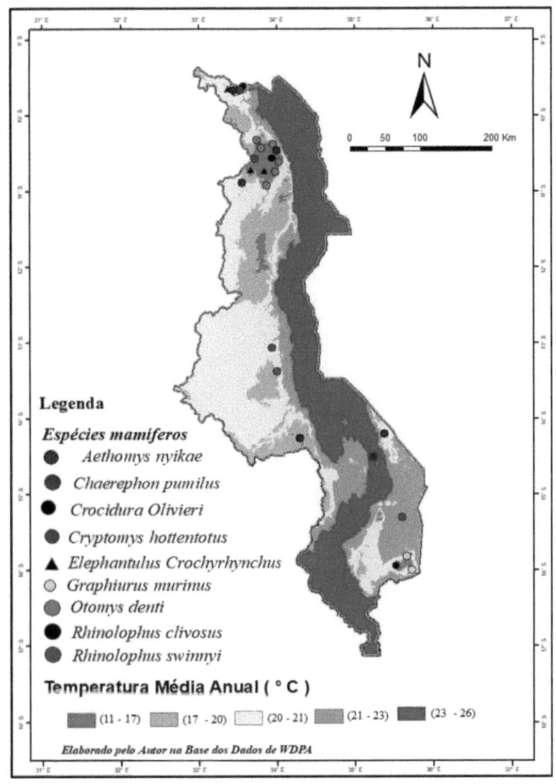

Figure 5 - Annual temperature variation generated in Arcgis 10.5 (Malawi).

4.3.3. Tanzania

Tanzania has an average annual maximum temperature of 27.8°C and a minimum of -3.3° C. The lowest average annual temperature was recorded in the Aruxa, Quilimanjaro, and Manyara regions, with a minimum of -3 to 17° C, as this is the highest altitude region in the country, while the highest average annual temperature was recorded in the Iringa, Lindi,

Rukwa, Pwani, Rovuma, Mtwara, Tanga, and Morogoro regions, ranging from 24 to 28° C.

The occurrence records show that most species such as Crocidura mdumai, Beamys hindei, Colobus angolensis, Colobus guereza occur in the 17-20° C range; Aethomys kaiseri, Mus musculoides, Beamys hindei between 20-22° C. In the range 24-28° C, the distribution of mammals is greater, with species such as: Arvicanthis neumanni, Congosorex phillipsorum, Crocidura munissii, Graphiurus microtis, Hipposideros vittatus, Miniopterus natalensis, Mops bakarii, Neoromicia nana, Nycteris thebaica, Pelomys minor, Petrodromus tetradactylus, Rhinolophus deckenii and Rhynchocyon petersi in the high temperature regions.

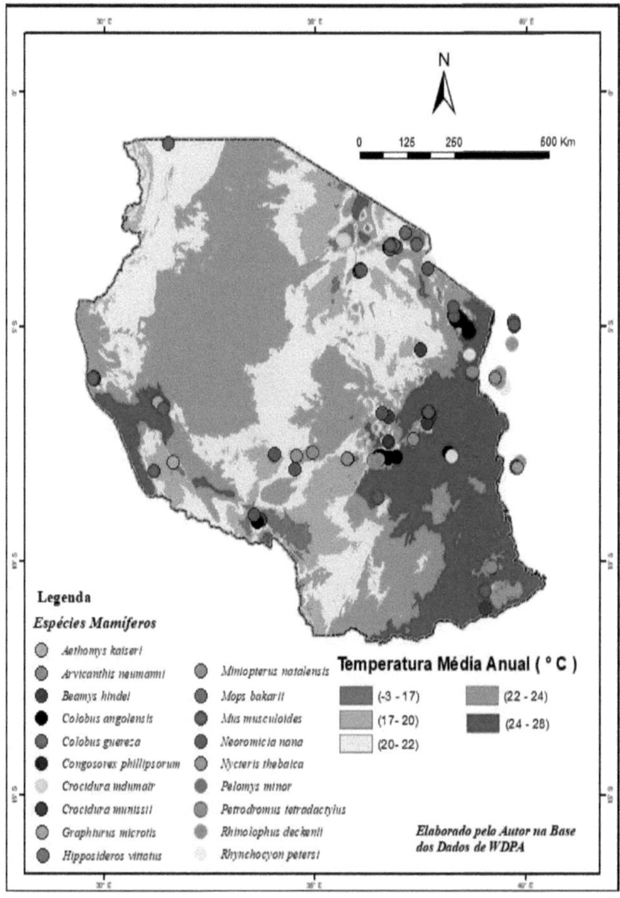

Figure 6 - Annual temperature variation generated in Arcgis 10.5 (Tanzania).

4.4. Influence of rainfall on the distribution of mammals

4.4.1. Mozambique

Mozambique has an average annual rainfall of 908 mm. The lowest average annual rainfall was recorded in the southern region (Gaza) with a range of 166-566 mm, while the highest average annual rainfall was recorded in the central region (Zambezia) ranging from 1,448 to 2,056 mm. Occurrence records show that the species Rhinolophus mossambicus occurs in the 566-855 mm range. In the 855 to 1,122 mm range, the species Nycteris thebaica, Epomophorus crypturus, Chaerephon pumilus and Rhinoloplus hildebrandt occur. These species are found in the central region, specifically in the buffer zone of Gorongosa National Park (PNG) and in the north, on Mozambique Island and, finally, between 1,122 and 1,448 mm, in the Gorongosa region, there is a very small group of Grammomys cometes (Figure 7).

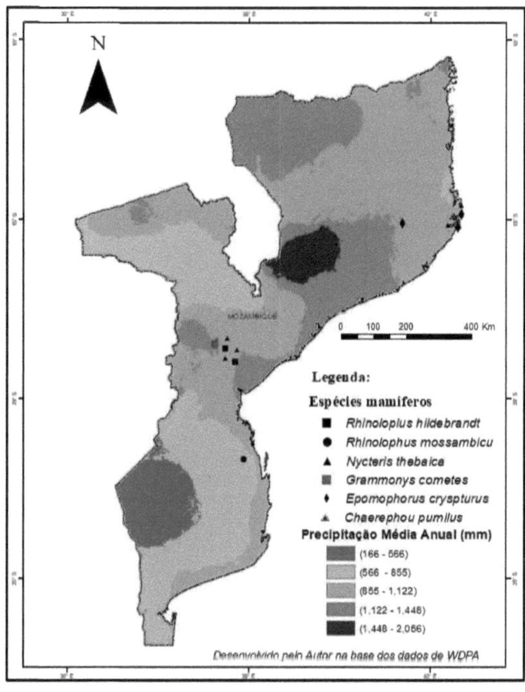

Figure 7 - Annual rainfall variation generated in Arcgis 10.5 (Mozambique).

4.4.2. Malawi

Malawi has an average annual rainfall of 1,068 mm. The lowest average annual rainfall was recorded in the North (Mzimba and Rumphi), Centre (Kasungu, Mchinji, Dowa and Lilongwe) and South (Chikwawa, Nsanje, Blantyre and Balaka), ranging from 724 to 926 mm, while the highest average annual rainfall was recorded in the North (Karonga and Nkhata Bay) and South (Mulanje and Thyolo) with a range of 1,530-2,496 mm. The occurrence records and distribution of the taxonomic group under study show that in the 724-926 mm range there is the species Aethomys nyikae, which is located in the Chongoni Forest Reserve. In the range between 926-1,078 mm are the species Chaerephon pumilus, and Cryptomys hottentotus located in Misuku Hills, Misuku Town, Mwalingo and Rhinolophus swinnyi from the Mt Mulanje region, Likhubula Camp. In the 1,078-1,273 mm range are the species Chaerephon pumilus, Crocidura olivieri, Rhinolophus clivosus, Elephantulus brachyrhynchus, Otomys denti. In the 1,273-1,530 mm range, the species Crocidura olivieri, Elephantulus brachyrhynchus, Rhinolophus clivosus, and Graphiurus murinus are present. In the higher rainfall range (1,530-2,496 mm) species such as Crocidura olivieri, Rhinolophus clivosus and Graphiurus murinus occur in the Ntchisi Forest Reserve.

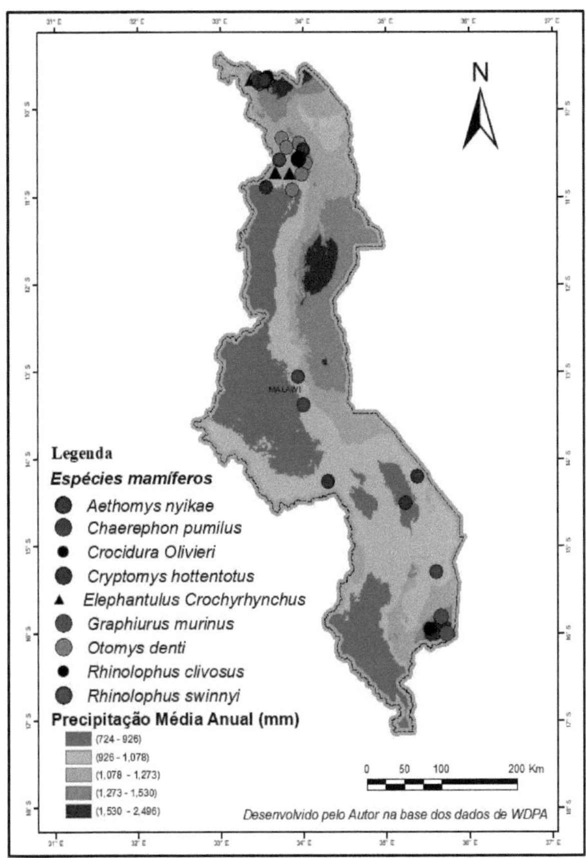

Figure 8 - Annual rainfall variation generated in Arcgis 10.5 (Malawi).

4.4.3. Tanzania

Tanzania has an average annual rainfall of 961 mm, with the lowest average annual variation recorded in the regions of Aruxa, Kilimanjaro, and Manyara, Tanga, Sangida and Dodoma. This can be explained by the topography of the region. The highest average annual rainfall was recorded in Kagera, Mwanza in the cross-border region between Tanzazina, Uganda and Kenya in the vicinity of Lake Victoria. The taxonomic group's occurrence records show that they are concentrated in the low rainfall intervals between 310-776 mm; Beamys hindei, Crocidura munissii, Neoromicia nana, Rhinolophus deckenii, Colobus guereza, Colobus angolensis and in the 776-1.000 mm, Mus musculoides, Aethomys kaiseri, Beamys hindei, Congosorex phillipsorum, Petrodromus tetradactylus and Crocidura mdumai; in the 1,250-1,578 mm range, Mus musculoides, Arvicanthis neumanni and Colobus angolensis and finally in the 1.2568-2,509 mm, Colobus angolensis, Colobus guereza, Beamys hindei and Mus musculoides are found with a small number of occurrences.

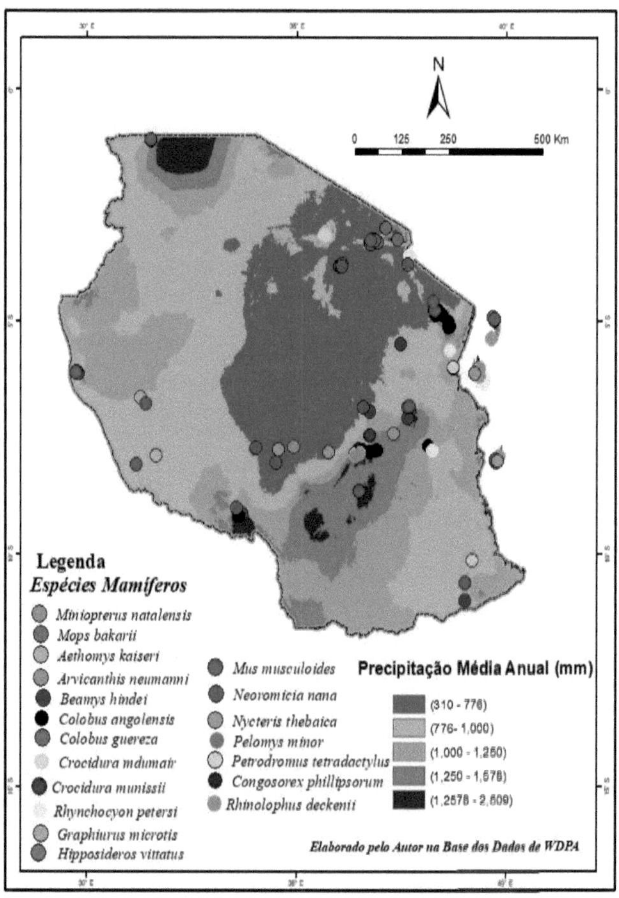

Figure 9 - Annual rainfall variation generated in Arcgis 10.5 (Tanzania).

4.5. The spatial distribution of mammals in relation to conservation units

4.5.1. Mozambique

Mozambique has 58 protected areas in its territory, which make up around 29.48 per cent of the coverage of terrestrial and inland water protected areas. Six (6) mammal species were recorded in this study: Chaerephon pumilus, Epomophorus crypturus, Grammomys cometes, Nycteris thebaica, Rhinolophus hildebrandtii and Rhinolophus mossambicus. These were mostly recorded in Gorongosa National Park, with four (4) species such as Rhinolophus mossambicus, Grammomys cometes, Nycteris thebaica and Epomophorus crypturus being recorded outside the Park. This result is a clear indicator of the spatial bias that characterises knowledge about mammal species, especially small mammals, in Mozambican territory.

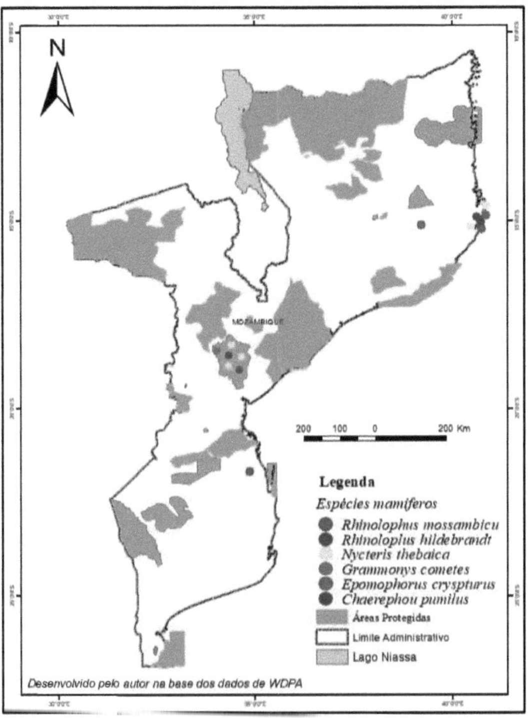

Figure 10 - Distribution of mammals in relation to Conservation Units (CUs) in Mozambique.

4.5.2. Malawi

Malawi has 133 protected areas in its territory, which make up around 22.88 per cent of the coverage of the country's terrestrial and inland water protected areas. Nine (9) mammal species were recorded in this study: Chaerephon pumilus, Aethomys nyikae, Crocidura olivieri, Cryptomys hottentotus, Elephantulus brachyrhynchus, Graphiurus murinus, Otomys denti, Rhinolophus clivosus, Rhinolophus swinnyi. These were mostly recorded in Malawi Parks and Reserves, with only three (3) species such as Aethomys nyikae, Rhinolophus swinnyi and Otomys denti recorded outside the Park. As in Mozambique, this result is an indicator of the spatial bias that characterises knowledge about mammal species, especially small mammals, in Malawian territory.

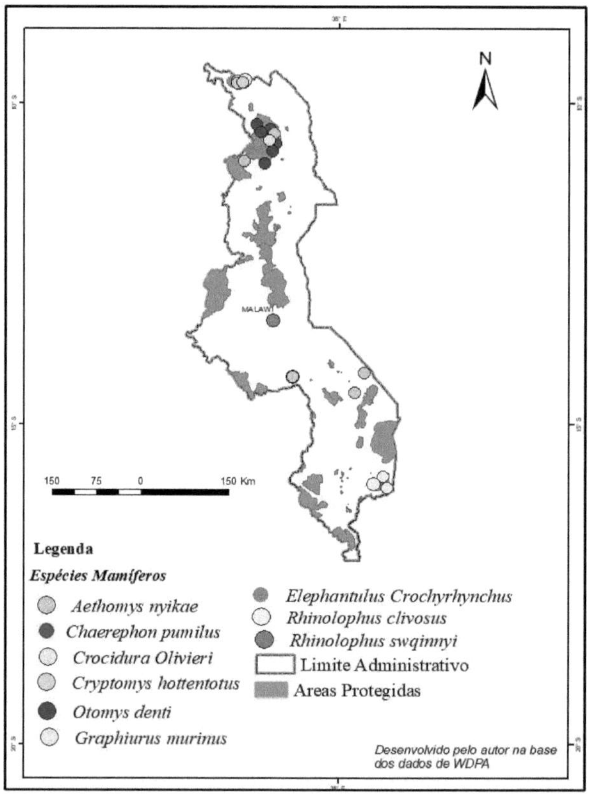

Figure 11 - Distribution of mammals in relation to Conservation Units (CUs) in Malawi.

5.5.3. Tanzania

Tanzania has 872 protected areas in its territory, which make up around 39.94 per cent of the coverage of the country's terrestrial and inland water protected areas. Nineteen (19) mammal species were recorded in this study: Aethomys kaiseri, Arvicanthis neumanni, Beamys hindei, Colobus angolensis, Colobus guereza, Congosorex phillipsorum, Crocidura mdumai, Crocidura munissii, Graphiurus microtis, Hipposideros vittatus, Miniopterus natalensis, Mops bakarii, Mus musculoides, Neoromicia nana, Nycteris thebaica, Pelomys minor, Petrodromus tetradactylus and Rhinolophus deckenii, of which six (6) species, Congosorex phillipsorum, Mops bakari, Mus musculoides, Crocidura mdumai, Rhynchocyon petersi and Aethomys kaiser, were recorded outside the protected areas. This result is once again an indicator of the spatial bias that characterises knowledge about small mammal species.

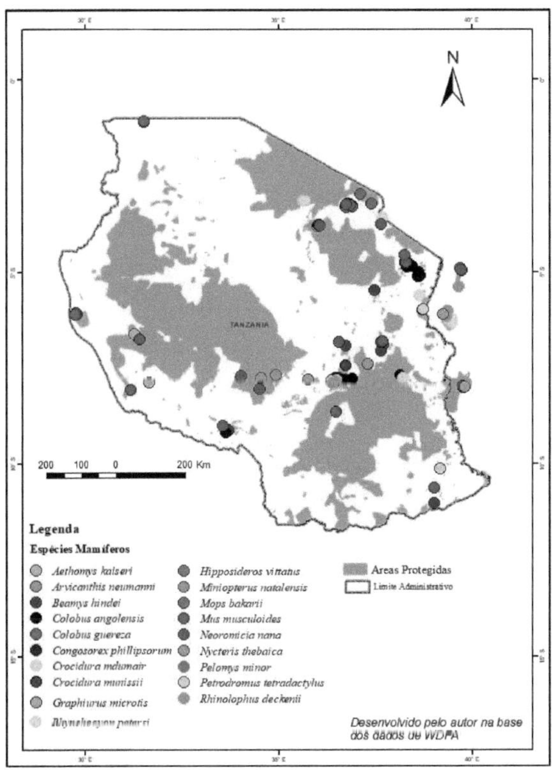

Figure 12 - Distribution of mammals in relation to Conservation Units (CUs) in Tanzania.

5.1. Evaluation of wealth

Spatial patterns of species richness, such as the latitudinal gradient of biodiversity, are an important indicator that can help in the study, conservation and management of wild species. Reliable description of spatial patterns of richness requires the most accurate knowledge possible of species distribution. However, current knowledge of the geographical distribution of most species is largely incomplete, especially in regions that are difficult and expensive to access (Menegotto & Rangel, 2018).

Given that the gaps in knowledge about species distribution are not randomly distributed in space, but are clustered in poorly sampled regions, the spatial patterns of species richness are statistically biased. Although analytical techniques have been developed to estimate the total richness of species in a community or region, these techniques generally do not take into account sampling effort and bias. This work sends a clear message that, regardless of the extraordinary value of sampling carried out in iconic areas such as Gorongosa National Park or Serengeti National Park, the description of biodiversity patterns, as well as the study of the variables that condition them, depend on work carried out in areas with a chronic lack of data and representative of the reality of the territory in terms of the environment and human pressure. The study area showed different diversity values between Mozambique, Malawi and Tanzania, with Tanzania showing a value of 3.03 and 0.95 for the Shannon-Weaver diversity index (H`) and Simpson's concentration index (C'), respectively, indicating that it is an area with relatively high diversity and a low concentration of species. The values for species richness, Shannon Index (H') and Simpson Index differ basically in the weight given to rare and abundant species. In the case of the Shannon Index, the weight given to rare and abundant species is identical. In the case of the Simpson Index, the weight given to rare species is lower. We can generalise this idea using diversity profiles (Melo, 2008). One of the great difficulties in comparing species richness (number of species) between communities stems from the difference in sampling effort (e.g. difference in the number of individuals, discrepancy in the number of sampling units or area sampled) which will inevitably influence the number of species observed (Michael Roswell, 2021). The rarefaction method allows us to compare the number of species between communities when the sample size (e.g., number of sampling units), sampling effort (e.g., sampling time) or number of individuals are not equal. Rarefaction calculates the expected number of species in each community based on a comparative value in which all the samples reach a standard size.

5.2. Influence of environmental variables

The visual and descriptive analyses of species distribution highlight the role of climate in defining the distribution patterns of organisms such as rodents, insectivores, bats and other species that deserved attention in this study, such as the two primates of the Cercopithecidae family found in Mount Kilimanjaro National Park (PNMK) and Arusha National Park (PNA) in Tanzania. The species identified as less tolerant are therefore species that should be prioritised for regular monitoring. The conclusions to be drawn are, however, limited, as the real impact of climate, land cover and human activities requires more spatially and temporally representative data.

The study by Schoeman & Monadjem (2018) concluded that reliable information on the taxonomic status of populations is fundamental for establishing priorities and making decisions for the conservation of species. Primates represent one of the best-studied groups of mammals in Africa. However, this taxonomic group is being debated and the geographical range, abundance and conservation status of most primates remain poorly known. This is especially the case in recent decades, given the rapid degradation, loss and fragmentation of primate habitats and the increase in hunting.

For the second largest taxonomic group in this study, bats, water availability and temperature are limiting factors (Mccain, 2007). This finding is also in line with the meta-analysis carried out by the same author, who concluded that for mountains permanently wet, the richness and diversity of bats is lower in the higher parts of the mountain. The distribution pattern of rodents in terms of elevation was different at the three sampling sites, and there seems to be a continuous decline in rodent richness with increasing elevation. The pattern may be different because forest ecosystems provide suitable habitat for other rodent species that tend to be less abundant with increasing altitude (Kafash et al., 2021). In addition to the uneven profiles of forest formations along the region's elevation gradient, another reasonable cause for the downward trend in diversity and richness is the variation in temperature at different altitudes in the study region. The temperature and altitude variables are directly related. Biological diversity includes all species on earth, all forms, levels and combinations of natural variation. If we are concerned about conserving biodiversity, we must understand how the number of species in a given place is determined and what factors regulate it. Species are not evenly distributed. The characteristics of the environment (temperature, rainfall), biotic factors (predation, competition) and human activities limit the distribution patterns of species.

5.3. Jaccard similarity index

Using the Jaccard similarity index (SJ), it was possible to find a relationship between the total number of mammal species found in the sampling region. The differences found indicate that the species are characterised by a non-uniform and heterogeneous distribution. As reported, heterogeneity must be taken into account when looking for ways to conserve the remaining biological diversity (Ferreira et al., 2008).

5.4. The spatial distribution of mammals in relation to conservation units

Thirty-four species of mammals georeferenced in the north-eastern region of southern Africa were used in this work. A range in the number of records was considered so that statistical and modelling analyses could be carried out. Although these were not carried out due to aspects related to the distribution and representativeness of the sampling, the distribution of the selected species in relation to the conservation units was assessed. Southern Africa is a region with very specific geophysical and ecological characteristics that limit the survival of some organisms that occur in the region, so information on the resources necessary for the survival of animals is also an important factor in understanding the distribution pattern of the group under study. The processes that determine the availability and distribution of these essential resources operate on different scales. Global scale (e.g., latitudinal and seasonal variation), meso-scale (e.g., climatic and elevation factors that control precipitation and radiation conditions), top-scale (e.g., terrain attributes, such as slope, that influence water and energy sources in the landscape), micro-scale (e.g., influence of the forest canopy on understory conditions) and nano-scale (e.g., soil and nutrient cycling by microorganisms) (Mackey & Lindenmayer, 2001).

Knowledge of the spatial distribution of species and the dynamics of populations in the landscape are important prerequisites for allocating conservation efforts, both at a large scale (global priorities) and at local and regional scales (national priorities) (Begon et al., 2006). The relative importance of scale measures on the distribution of organisms varies according to the species, due to differences related to the biology and habitat requirements of the species itself (Gutzwiller, 2002). Eastern and Southern Africa has 5,232 protected areas covering an area of 2,120,112 km^2 . The data was calculated using the spatial information available in the WDPA, in combination with a number of other authoritative datasets. However, in some cases, the boundaries of protected areas in the WDPA are not up to date, which can affect the accuracy of the statistics for each country or region (IUCN, 2021).

Maps should therefore be used with as much care and judgement as possible, always bearing in mind their limitations. Similarly, sites with old presence records and no recent records for a particular species do not necessarily mean that the species has disappeared from the site, but only that a more recent record of its presence has not been obtained, often due to a lack of prospecting in the field. The available data is therefore not sufficient to infer temporal trends in the distribution of the taxonomic group selected in this study.

This study revealed deficits in the information needed to manage, monitor and apply specific measures used to conserve biodiversity, especially small mammals (Neves et al., 2019). The plans are more focussed on large mammals and there are annual reports on the management of these species. The role of protected areas (PAs) has been widely quantified, with an emphasis on preserving the integrity of ecosystems and biodiversity. On the other hand, the ecological impact of protected areas (PAs) on wildlife, in terms of their effects on the ecological niche of species, has rarely been studied. This research gap is particularly pertinent given the low representation of the vertebrate species niche in the global network of protected areas (PAs) and the strong ecological contrast between protected and unprotected lands worldwide (Santangeli et al., 2022). This was one of the gaps found in this study.

The same author's study reveals that all these species characteristics can thus mediate the differences in ecological niche properties between protected and non-protected areas. In relation to the conservation status of species, we predict that threatened species, which are typically more dependent on protected areas and the ecosystems within them, may show less change in niche properties at the interface between protected and unprotected areas. On the other hand, non-threatened species may be more adaptable and adjust their niche properties according to the different environmental conditions present inside and outside protected areas (Santangeli et al., 2022). Many of these species probably did not persist in areas that are now protected by legislation in their countries, and it is to be hoped that current conservation efforts, although still modest, will help them to recover in distribution and abundance. It is now necessary to move forward, increasing conservation efforts for these threatened species, while at the same time carrying out new surveys to obtain a more complete assessment of the diversity of the region under study and the conservation needs of the entire mammal fauna. Geographical distribution models consider that if a species is found in certain conditions, it can survive and reproduce in other places with the same characteristics, so the validation of environmental favourability and potential distribution maps can be considered a useful tool from an ecological point of view, but also for the conservation and management of the species studied. To this end, the proper collection and collation of information is essential.

6. CONCLUSIONS AND SUGGESTIONS

This research has provided a critical analysis of the distribution of small mammals in the north-eastern region of southern Africa. In addition, the study shows that sampling efforts are centred on protected areas, which makes it impossible to study the environmental and anthropogenic variables that in some way influence the presence and conservation of mammals. Given the context of most protected areas in Africa, small mammals have received less attention compared to large mammals, even though they play a fundamental role in maintaining the functionality of ecosystems. It is therefore necessary to understand how small mammals are distributed in a given ecosystem as baseline information to enable holistic and informed management (Saanya et al., 2022). The data collated in this study demonstrates the lack of information on the distribution of mammals in the countries considered. Despite recent meritorious efforts (Neves et al., 2018), much remains to be done to make this information more spatially representative. This information is essential to ensure the adoption of good conservation practices for biodiversity.

REFERENCES BIBLIOGRAPHIC

Amori, G., Masciola, S., Saarto, J., Gippoliti, S., Rondinini, C., Chiozza, F., & Luiselli, L. (2012). Spatial turnover and knowledge gap of African small mammals: Using country checklists as a conservation tool. Biodiversity and Conservation, 21(7), 1755-1793. https://doi.org/10.1007/s10531-012-0275-5

Balčiauskas, L., & Balčiauskienė, L. (2022). Small Mammal Diversity Changes in a Baltic Country, 1975-2021: A Review. Life, 12(11). https://doi.org/10.3390/life12111887

Begon, M.; Harper, J.L.; Townsend, C.R. (2006). Ecology: from individuals to ecosystems. 4th. ed. Malden: Blackwell Publishing, 738 p.

Boitani, L., Maiorano, L., Baisero, D., Falcucci, A., Visconti, P., & Rondinini, C. (2011). What spatial data do we need to develop global mammal conservation strategies? Philosophical Transactions of the Royal Society B: Biological Sciences, 366(1578), 2623-2632. https://doi.org/10.1098/rstb.2011.0117

Catullo, G., Masi, M., Falcucci, A., Maiorano, L., Rondinini, C., & Boitani, L. (2008). A gap analysis of Southeast Asian mammals based on habitat suitability models. Biological Conservation, 141(11), 2730-2744. https://doi.org/10.1016/j.biocon.2008.08.019

Chen, Y., & Peng, S. (2017). Evidence and mapping of extinction debts for global forest-dwelling reptiles, amphibians and mammals. Scientific Reports, 7(March), 1-10. https://doi.org/10.1038/srep44305

D.C.D Happold, M. H. & J. E. H. (1987). The bats of Malawi. 3, 51. https://doi.org/doi:10.1515/mamm.1987.51.3.337

Dormann, C. F. (2007). Effects of incorporating spatial autocorrelation into the analysis of species distribution data. Global Ecology and Biogeography, 16(2), 129-138. https://doi.org/10.1111/j.1466-8238.2006.00279.x

Eklund, J., Arponen, A., Visconti, P., & Cabeza, M. (2011). Governance factors in the identification of global conservation priorities for mammals. Philosophical Transactions of the Royal Society B : Biological Sciences, 366(1578), 2661-2669. https://doi.org/10.1098/rstb.2011.0114

Ferreira, E. V., Soares, T. S., Da Costa, M. F. F., & Silva, V. S. M. (2008). Composition, diversity and floristic similarity of a submontane tropical semideciduous forest in Marcelândia - MT. Acta Amazonica, 38(4), 673-679. https://doi.org/10.1590/S0044- 59672008000400010

Frank, A. S. K., & Schäffler, L. (2019). Identifying key knowledge gaps to better protect biodiversity and simultaneously secure livelihoods in a priority conservation area. Sustainability (Switzerland), 11(20), 1-22. https://doi.org/10.3390/su11205695

Freitag, S., & Van Jaarsveld, A. S. (1998). Sensitivity of selection procedures for priority conservation areas to survey extent, survey intensity and taxonomic knowledge. Proceedings of the Royal Society B: Biological Sciences, 265(1405), 1475-1482. https://doi.org/10.1098/rspb.1998.0460

Fritz, S. A., Bininda-Emonds, O. R. P., & Purvis, A. (2009). Geographical variation in predictors of mammalian extinction risk: Big is bad, but only in the tropics. Ecology Letters, 12(6), 538- 549. https://doi.org/10.1111/j.1461-0248.2009.01307.x
Gutzwiller K.J. (2002). Applying landscape ecology in biological conservation. New York: Springer-Verlag. 518 p.

Huntley, B. J., Russo, V., Lages, F., & Ferrand, N. (2019). Biodiversity of angola: Science & conservation: A modern synthesis. In Biodiversity of Angola: Science and Conservation: A Modern Synthesis. https://doi.org/10.1007/978-3-030-03083-4

IUCN. (2021). State of protected and conservation areas in Eastern and Southern Africa. In U. ESARO (1st ed.), IUCN, Eastern and Southern Africa Regional Office (ESARO) Biodiversity and Protected Area Management (BIOPAMA) Programme. https://doi.org/10.2305/iucn.ch.2020.15.pt

Joppa, L. N., O'Connor, B., Visconti, P., Smith, C., Geldmann, J., Hoffmann, M., Watson, J. E. M., Butchart, S. H. M., Virah-Sawmy, M., Halpern, B. S., Ahmed, S. E., Balmford, A., Sutherland, W. J., Harfoot, M., Hilton-Taylor, C., Foden, W., Di Minin, E., Pagad, S., Genovesi, P., ... Burgess, N. D. (2016). Filling in biodiversity threat gaps: Only 5 per cent of global threat data sets meet a "gold standard." Science, 352(6284), 416-418. https://doi.org/10.1126/science.aaf3565

Karim, R., & Ahsan, F. (2016). Mammalian Fauna and Conservational Issues of the

Baraiyadhala National Park in Chittagong , Bangladesh. Open Journal of Forestry, 6(April), 123-134. https://doi.org/DOI: 10.4236/ojf.2016.62011

Laurance, W. F., Sayer, J., & Cassman, K. G. (2014). Agricultural expansion and its impacts on tropical nature. Trends in Ecology and Evolution, 29(2), 107-116. https://doi.org/10.1016/j.tree.2013.12.001

Lecours, V. (2017). On the use of maps and models in conservation and resource management (Warning: Results may vary). Frontiers in Marine Science, 4(SEP), 1-18. https://doi.org/10.3389/fmars.2017.00288

Mackey, B.G.; Lindenmayer, D.B. (2001) Towards a hierarchical framework for modelling the spatial distribution of animals. Journal of Biogeography, Oxford, v. 28, p. 1147-1166.

Masoud Yousefi, Ahmad Mahmoudi, Anooshe Kafash, A. K. and B. K. (2021). Biogeography of rodents in Iran: species richness, elevational distribution and their environmental correlates. Original Study – DE GRUYTER, 86(4): 309-320. https://doi.org/https://doi.org/10.1515/mammalia-2021-0104

Mccain, C. M. (2007). Could temperature and water availability drive elevational species richness patterns ? A global case study for bats. 1-13. https://doi.org/10.1111/j.1466-822x.2006.00263.x

Melo, A. S. (2008). What do we win "confounding" species richness and evenness in a diversity index? Biota Neotropica, 8(3), 21-27. https://doi.org/10.1590/s1676-06032008000300001

Menegotto, A., & Rangel, T. F. (2018). Mapping knowledge gaps in marine diversity reveals a latitudinal gradient of missing species richness. Nature Communications, 9(1). https://doi.org/10.1038/s41467-018-07217-7

Michael Roswell, J. D. and R. W. (2021). A conceptual guide to measuring species diversity.

Oikos, 130: 321-338. https://doi.org/doi: 10.1111/oik.07202

Naeem, S., Chazdon, R., Duffy, J. E., Prager, C., & Worm, B. (2016). Biodiversity and human well-being: An essential link for sustainable development. Proceedings of the Royal Society B: Biological Sciences, 283(1844). https://doi.org/10.1098/rspb.2016.2091

Neves, I. Q., da Luz Mathias, M., & Bastos-Silveira, C. (2018). The terrestrial mammals of Mozambique: Integrating dispersed biodiversity data. Bothalia, 48(1), 1-23.

https://doi.org/10.4102/ABC.V48I1.2330

Neves, I. Q., Mathias, M. da L., & Bastos-Silveira, C. (2019). Mapping Knowledge Gaps of Mozambique's Terrestrial Mammals. Scientific Reports, 9(1), 1-14. https://doi.org/10.1038/s41598-019-54590-4

Peterson, A. T., & Soberón, J. (2012). Species distribution modelling and ecological niche modelling: Getting the Concepts Right. Nature to Conservation, 10(2), 102-107. https://doi.org/10.4322/natcon.2012.019

Rondinini, C., Stuart, S., & Boitani, L. (2005). Habitat suitability models and the shortfall in conservation planning for African vertebrates. Conservation Biology, 19(5), 1488-1497. https://doi.org/10.1111/j.1523-1739.2005.00204.x

Rondinini, C., Wilson, K. A., Boitani, L., Grantham, H., & Possingham, H. P. (2006). Tradeoffs of different types of species occurrence data for use in systematic conservation planning. Ecology Letters, 9(10), 1136-1145. https://doi.org/10.1111/j.1461-0248.2006.00970.x

Saanya, A., Massawe, A., & Makundi, R. (2022). Small Mammal Species Diversity and Distribution in the Selous Ecosystem, Tanzania. African Zoology, 57(1), 20-31. https://doi.org/10.1080/15627020.2022.2034040

Santangeli, A., Mammola, S., Lehikoinen, A., Rajasärkkä, A., Lindén, A., & Saastamoinen, M. (2022). The effects of protected areas on the ecological niches of birds and mammals. Scientific Reports, 12(1), 1-12. https://doi.org/10.1038/s41598-022-15949-2

Schipper, J., Chanson, J. S., Chiozza, F., Cox, N. A., Hoffmann, M., Katariya, V., Lamoreux, J., Rodrigues, A. S. L., Stuart, S. N., Temple, H. J., Baillie, J., Boitani, L., Lacher, T. E., Mittermeier, R. A., Smith, A. T., Absolon, D., Aguiar, J. M., Amori, G., Bakkour, N., ... Young, B. E. (2008). The status of the world's land and marine mammals: diversity, threat, and knowledge. Science, 322(5899), 225-230. https://doi.org/10.1126/science.1165115

Schmid, R., Wilson, D. E., & Reeder, D. M. (1993). Mammal Species of the World: A Taxonomic and Geographic Reference. Taxon, 42(2), 512. https://doi.org/10.2307/1223169

Schoeman, M. C., & Monadjem, A. (2018). Community structure of bats in the savannas of southern Africa: influence of scale and human land-use. Hystrix, Italian Journal of Mammalogy, 29(1), 3-10. https://doi.org/10.4404/hystrix

Uramoto, K., Walder, J. M. M., & Zucchi, R. A. (2005). Quantitative analysis and distribution of populations of anastrepha species (Diptera: Tephritidae) on the Luiz de Queiroz Campus, Piracicaba, SP. Neotropical Entomology, 34(1), 33-39. https://doi.org/10.1590/S1519-566X2005000100005

Wilson, K. A., Evans, M. C., di Marco, M., Green, D. C., Boitani, L., Possingham, H. P., Chiozza, F., & Rondinini, C. (2011). Prioritising conservation investments for mammal species globally. Philosophical Transactions of the Royal Society B: Biological Sciences, 366(1578), 2670-2680. https://doi.org/10.1098/rstb.2011.0108

Mozambique, UNEP-WCMC and IUCN (2022), Protected Planet: The World Database on Protected Areas (WDPA) and World Database on Other Effective Area-based Conservation Measures (WD-OECM) [Online], July 2022, Cambridge, UK: UNEP-WCMC and IUCN. Available at: www.protectedplanet.net.

Malawi, UNEP-WCMC and IUCN (2022), Protected Planet: The World Database on Protected Areas (WDPA) and World Database on Other Effective Area-based Conservation Measures (WD-OECM) [Online], July 2022, Cambridge, UK: UNEP-WCMC and IUCN. Available at: www.protectedplanet.net.

Tanzania, UNEP-WCMC and IUCN (2022), Protected Planet: The World Database on Protected Areas (WDPA) and World Database on Other Effective Area-based Conservation Measures (WD-OECM) [Online], July 2022, Cambridge, UK: UNEP-WCMC and IUCN. Available at: www.protectedplanet.net.

GBIF (2011), Global Biodiversity Information Facility. See (accessed 15 March 2011). GBIF (2021), http://www.gbif.org. (5 January 2021) The Global Biodiversity Information Facility,with the reference https://doi.org/10.15468/dl.52f654 UCN Red List (2011), The International Union for the Conservation of Nature's Red List of Threatened Species. See http://www.iucnredlist.org/initiatives/mammals (accessed 15 March 2011).

APPENDIXES

Appendix 1 - Conservation status of the species under study according to an exploratory analysis of the IUCN database, 2022.

Ord.	Distribution Area	Family	Species	Conservation status	List category and criteria IUCN red
1	MOZ	Molossidae	Chaerephon pumilus	No information that it is present in some protected areas.	Of Little Concern (PP)
2	MOZ	Pteropodidae	Epomophorus crypturus	It occurs in many protected areas.	Of Little Concern (PP)
3	MOZ	Muridae	Grammomys cometes	No taxonomic information. Populations of this species should be monitored to record changes in its abundance and distribution.	Not much Worrying (PP)
4	MOZ	Nycteridae	Nycteris thebaica	It may occur in some protected areas. No specific conservation measures are known.	Not much Worrying (PP)
5	MOZ	Rhinolophidae	Rhinolophus hildebrandtii	There seem to be no direct conservation measures in vigour. The species is unlikely to be present in some protected areas	Not much Worrying (PP)
6	MOZ	Rhinolophidae	Rhinolophus mossambicus	It occurs in the Niassa Reserve and Gorongosa National Park. Needs to be updated to understand its distribution and possible threats.	Not much Worrying (PP)
7	MW	Molossidae	Chaerephon pumilus	This species is unlikely to be present in some protected areas.	Of Little Concern (PP)

	MW	Muridae	Aethomys nyikae	There is no concrete information about the species.	Of Little Concern (PP)
9	MW	Soricidae	Crocidura olivieri	It was collected in various protected areas.	Little Worrying (PP)
10	MW	Bathyergidae	Cryptomys hottentotus	The species occurs in many protected areas throughout its range.	Of Little Concern (PP)
11	MW	Macroscelididae	Elephantulus brachyrhynchus	The species occurs in protected areas, but there is no information on its distribution.	Not much Worrying (PP)
12	MW	Gliridae	Graphiurus murinus	It is presumed to be present in some protected areas.	Not much Worrying (PP)
13	MW	Muridae	Otomys denti	They are located within protected areas.	Of Little Concern (PP)
14	MW	Rhinolophidae	Rhinolophus clivosus	It is likely to be present in a number of protected areas. Because it is a group of migratory species.	Not much Worrying (PP)
15	MW	Rhinolophidae	Rhinolophus swinnyi	It is present in some protected areas (e.g. e.g. Kruger National Park, South Africa). Little information on its distribution in the MW.	Not much Worrying (PP)
16	TZ	Muridae	Aethomys kaiseri	There are no conservation measures in place. It is not known if the species is present in any areas protected	Of Little Concern (PP)
17	TZ	Muridae	Arvicanthis neumanni	There are no conservation measures in place. No it is not known whether the species is present in any protected areas.	Not much Worrying (PP)
18	TZ	Nesomyidae	Beamys hindei	It is found in Udzungwa National Park and in some forest reserves. It occurs in some areas protected in Malawi.	Of Little Concern (PP)

19	TZ	Cercopith ecidae	Colobus	It occurs in the mountains of Tanzania's Eastern Arc,	
			angolensis	in many protected areas is classified as Endangered and Vulnerable due to its small distribution area.	Vulnerable (VU)
20	TZ	Cercopith ecidae	Colobus guereza	Protected in two national parks: Mount Kilimanjaro National Park and Arusha National Park in Tanzania).	Not much Worrying (PP)
21	TZ	Soricidae	Congosore x phillipsoru m	This species has not been updated in any protected area, but may be present in the Park Udzungwa National Park.	Critically Endangered (CR),
22	TZ	Soricidae	Crocidura mdumai	This species occurs in several protected areas.	Not Evaluated (NA)
23	TZ	Soricidae	Crocidura munissii	It occurs predominantly in protected areas.	Of Little Concern (PP)
24	TZ	Gliridae	Graphiuru s microtis	This species occurs in several protected areas.	Little Worrying (PP)
25	TZ	Hipposide ridae	Hipposide ros vittatus	This species is present in several protected areas in East Africa. In southern Africa, it has been recorded in the Kruger National Park, in the province of	Near Threatened (QA)
				Limpopo, South Africa, and Limpopo National Park. Gorongosa.	
26	TZ	Miniopteri dae	Miniopter us natalensis	Apparently, there are no conservation measures in place for this species.	Not much Worrying (PP)
27	TZ	Molossida e	Mops bakarii	There are no specific conservation measures for this species.	Information Insufficient (II)
28	TZ	Muridae	Mus musculold es	It is presumed to be present in many protected areas.	Not much Worrying (PP)

29	TZ	Vespertili onidae	Neoromici a nana	It is probably present in many protected areas. More research is needed on the taxonomy of this species.	Of Little Concern (PP)
30	TZ	Nycterida e	Nycteris thebaica	It can occur in some protected areas. É More research is needed into the size and trends of this population.	Not much Worrying (PP)
31	TZ	Muridae	Pelomys minor	There is no concrete information on this species. It is not known whether the species is present in some protected area.	Of Little Concern (PP)
32	TZ	Macroscel ididae	Petrodrom us tetradactyl us	The species occurs in protected areas, but these have not been counted or assessed.	Of Little Concern (PP)
33	TZ	Rhinoloph idae	Rhinoloph us deckenii	There is no concrete information about this species, but is present in some protected areas, the RF and the greater Gorongosa.	Near Threatened (QA)
34	TZ	Macroscel ididae	Rhynchoc yon petersi	Occurring in numerous relatively small forest blocks, they are referred to as forest reserves in Kenya and Tanzania.	Of Little Concern (PP)

Appendix 2 - Occurrence of species by location and country.

Species by location	Mozambique	Malawi	Tanzania	Total
1.6 km SE (by air) Chinunka		4		4
Elephantulus brachyrhynchus		4		4
Amboni Caves			23	23
Triaenops afer			23	23
Blantyre, Top Mandala Museum		1		1
Crocidura olivieri		1		1
Bujingijila corridor between Mt Rungwe and Livingstone Mts			5	5
Beamys hindei			1	1
Suncus megalura			4	4
Chihalatan Caves (Gerhard's Cave), 38.2 km E Inhassoro	34			34
Nycteris thebaica	14			14
Rhinolophus mossambicus	20			20
Chipata Hill, Nkhotakota Wildlife Reserve		6		6
Rhinolophus swinnyi		6		6
Chongoni Forest Reserve		3		3
Aethomys nyikae		2		2
Rhinolophus clivosus		1		1
Emao			2	2
Mus musculoides			2	2
Gendagenda Forest			1	1
Rhynchocyon petersi			1	1
Gorongosa National Park, montane forest	16			16
Grammomys cometes	16			16
Gorongosa National Park, Muaredzi River, karst cave	15			15
Nycteris thebaica	1			1
Rhinolophus hildebrandtii	14			14
Gorongosa National Park; Bela Vista along the Pungwe River	5			5
Rhinolophus hildebrandtii	5			5
Gorongosa National Park; Bunga Inselberg camp	1			1

Nycteris thebaica	1		1
Gorongosa National Park; Bunga Inselberg grotto	3		3
Nycteris thebaica	3		3
Lumbo Catholic Church	12		12
Chaerephon pumilus	12		12
Mozambique Island	7		7
Epomophorus crypturus	7		7
Ilha de Mozambique, in abandoned building	5		5
Epomophorus crypturus	5		5
Mozambique Island, Museum of Mozambique Island	6		6
Nycteris thebaica	6		6
Mozambique Island, near Hindu Temple	5		5
Chaerephon pumilus	5		5
Juani Island		4	4
Nycteris thebaica		2	2
Rhinolophus deckenii		2	2
Juani Island, Juani village		1	1
Rhinolophus deckenii		1	1
Juani Island, Kua ruins		1	1
Neoromicia nana		1	1
Kambona Forest Reserve		3	3
Beamys hindei		2	2
Mus musculoides		1	1
Kibebe Farms, approx 6 km ESE Iringa		5	5
Aethomys kaiseri		1	1
Arvicanthis neumanni		2	2
Mus musculoides		1	1
Nycteris thebaica		1	1
Kichangani		1	1
Mus musculoides		1	1
Kijijini village		21	21
Hipposideros vittatus		21	21

Kilakala			1	1
Mus musculoides			1	1
Kilombero Valley			1	1
Mus musculoides			1	1
Kissona village	1			1
Nycteris thebaica	1			1
Kola Hill			1	1
Mus musculoides			1	1
Likhubula Station, Mt. Mulanje		1		1
Crocidura olivieri		1		1
Loggers Camp, Katavi National Park, Mpanda District, Rukwa Region, Western Tanzania			1	1
Aethomys kaiseri			1	1
Mafia Island, Mlola Forest			4	4
Rhinolophus deckenii			4	4
Mafia Island, Utende			2	2
Neoromicia nana			1	1
Nycteris thebaica			1	1
Mafiga			1	1
Mus musculoides			1	1
Magamba			1	1
Mus musculoides			1	1
Mahale Mts, Mahale National Park, 0.5 km NW Nkungwe Hill summit			4	4
Mus musculoides			1	1
Suncus megalura			3	3
Mahale Mts, Mahale National Park, 0.5 km S Pasagulu Hill			2	2
Mus musculoides			2	2
Mahale Mts, Mahale National Park, edge of Lake Tanganyika			2	2
Mus musculoides			2	2
Mangochi Hills, Skull Rock Scout Camp		4		4
Aethomys nyikae		4		4
Mbizi Mts, 1 km S, 1 km E Wipanga			7	7
Aethomys kaiseri			7	7

Mbizi Mts, Mbizi Forest Reserve, 1/2 km S, 3 km E Wipanga		8	8
Aethomys kaiseri		8	8
Mbuyuni village, Maruwa Jongo cave		2	2
Rhinolophus deckenii		2	2
Meru		1	1
Mus musculoides		1	1
Mikumi National Park, Malundwe Mountains		16	16
Beamys hindei		4	4
Mus musculoides		2	2
Nycteris thebaica		1	1
Rhinolophus deckenii		9	9
Minziro Forest		14	14
Mus musculoides		1	1
Neoromicia nana		7	7
Nycteris thebaica		1	1
Suncus megalura		5	5
Misuku Hills, Misuku Town, Mwalingo	36		36
Chaerephon pumilus	16		16
Crocidura olivieri	1		1
Cryptomys hottentotus	17		17
Graphiurus murinus	2		2
Misuku Hills, Mughese Forest	23		23
Aethomys nyikae	1		1
Crocidura olivieri	6		6
Graphiurus murinus	15		15
Rhinolophus clivosus	1		1
Misuku Hills, Wilindi Forest	12		12
Crocidura olivieri	1		1
Graphiurus murinus	11		11
Mkunya River Forest Reserve		1	1
Beamys hindei		1	1
Mpanda		3	3

Mus musculoides			3	3	
Mt Mulanje, Chisongeli Forest		3		3	
Crocidura olivieri		2		2	
Graphiurus murinus		1		1	
Mt Mulanje, Likhubula Camp		1		1	
Rhinolophus swinnyi		1		1	
Mt Mulanje, Mulanje Forest Res, edge Lujeri Tea Estate, next to Ruo River		1		1	
Graphiurus murinus		1		1	
Mt Mulanje, Rua River Valley, above Lujeri Tea Estate		2		2	
Crocidura olivieri		2		2	
Mt Rungwe, Rungwe Nature Reserve			1	1	
Colobus angolensis			1	1	
Mulanje Forest Reserve, Mt. Mulanje		3		3	
Rhinolophus clivosus		3		3	
Namizimu Forest Reserve, Kwitunji Camp		1		1	
Aethomys nyikae		1		1	
Nampula, outskirts of town on road towards Malawi, Bamboo Hotel	3			3	
Epomophorus crypturus	3			3	
Ngezi Forest			1	1	
Neoromicia nana			1	1	
Ngezi Forest, Kipangani village			21	21	
Mops bakarii			21	21	
Ngorongoro Conservation Area, Ngorongoro Crater rim, near Lamala Gate			6	6	
Crocidura mdumai			6	6	
Ngorongoro Conservation Area, Ngorongoro Crater rim, near Pongo Ranger Post			11	11	
Crocidura mdumai			11	11	
Nguu Mts, Nguru North Forest Reserve, 5.6 km S, 3 km E Gombero			2	2	
Beamys hindei			2	2	
North Pare Mts, Kindoroko Forest Reserve			1	1	
Neoromicia nana			1	1	
North Pare Mts, Minja Forest Reserve			1	1	
Rhynchocyon petersi			1	1	

Northern Region, North "A" District, Pashuni cave		1	1
Triaenops afer		1	1
Northern Region, North "B" District, Kiwengwa-Pongwe Forest Reserve cave		2	2
Rhinolophus deckenii		2	2
Ntantwa Forest		1	1
Mus musculoides		1	1
Ntchisi Forest Reserve	29		29
Crocidura olivieri	3		3
Rhinolophus clivosus	10		10
Rhinolophus swinnyi	16		16
Nyika National Park, 18 km N Thazima, Runyina Bridge	15		15
Aethomys nyikae	6		6
Elephantulus brachyrhynchus	9		9
Nyika National Park, below Chilinda Dam 2	2		2
Otomys denti	2		2
Nyika National Park, Chelinda Dam	7		7
Crocidura olivieri	1		1
Otomys denti	6		6
Nyika National Park, Chilinda Camp	7		7
Aethomys nyikae	2		2
Crocidura olivieri	1		1
Elephantulus brachyrhynchus	2		2
Otomys denti	2		2
Nyika National Park, Chilinda Camp, Dam 2	5		5
Crocidura olivieri	2		2
Otomys denti	3		3
Nyika National Park, Mwenembwe Forest	1		1
Otomys denti	1		1
Nyika National Park, Zovo Chipolo Forest	2		2
Otomys denti	1		1
Rhinolophus swinnyi	1		1
Poroto Mts, Ngozi Crater		1	1

Neoromicia nana	1	1
Rondo Forest Reserve	4	4
Beamys hindei	1	1
Petrodromus tetradactylus	3	3
Ruaha National Park, Isunkavyola Plateau	15	15
Beamys hindei	1	1
Pelomys minor	14	14
Ruaha National Park, Isunkavyola Plateau, Kilola Star	36	36
Beamys hindei	2	2
Miniopterus natalensis	21	21
Neoromicia nana	2	2
Pelomys minor	11	11
Ruaha National Park, Maji Moto	3	3
Mus musculoides	1	1
Neoromicia nana	1	1
Nycteris thebaica	1	1
Ruaha National Park, Mikindi Springs	4	4
Mus musculoides	1	1
Nycteris thebaica	3	3
Ruaha National Park, Msembe Headquarters, Dispensary building	22	22
Arvicanthis neumanni	22	22
Rubeho Mts, Ilole forest	12	12
Beamys hindei	4	4
Crocidura munissii	8	8
Rubeho Mts, Mwofwomero Forest Reserve	1	1
Beamys hindei	1	1
Rubeho Mts, Mwofwomero forest, near Chugu Peak	7	7
Beamys hindei	1	1
Crocidura munissii	3	3
Neoromicia nana	1	1
Rhinolophus deckenii	2	2
Southern Region, Central District, Kwabaniani caves	2	2

Rhinolophus deckenii	2	2
Tanzania - Arusha	1	1
Colobus guereza	1	1
Tanzania - Arusha NP	15	15
Colobus guereza	15	15
Tanzania - Handeni	1	1
Colobus angolensis	1	1
Tanzania - Jozani-Chwaka Bay NP	3	3
Rhynchocyon petersi	3	3
Tanzania - Kilimanjaro NP	1	1
Colobus guereza	1	1
Tanzania - Kilombero	1	1
Colobus angolensis	1	1
Tanzania - Korogwe	1	1
Colobus angolensis	1	1
Tanzania - Kusini	11	11
Rhynchocyon petersi	11	11
Tanzania - Lushoto	8	8
Colobus angolensis	8	8
Tanzania - Moshi Rural	1	1
Colobus guereza	1	1
Tanzania - Mount Meru NP	1	1
Colobus guereza	1	1
Tanzania - Muheza	3	3
Colobus angolensis	3	3
Tanzania - Rufiji	3	3
Colobus angolensis	1	1
Rhynchocyon petersi	2	2
Tanzania - Saadani NP	4	4
Colobus angolensis	1	1
Petrodromus tetradactylus	1	1
Rhynchocyon petersi	2	2

Tanzania - Udzungwa Mountains NP	2	2
Colobus angolensis	2	2
Tanzania - Zansibar Central	1	1
Rhynchocyon petersi	1	1
Tanzania, United Republic of	1	1
Colobus guereza	1	1
Tarangire National Park, Foley's Camp	9	9
Graphiurus microtis	8	8
Neoromicia nana	1	1
Tarangire National Park, Lemiyon area	5	5
Graphiurus microtis	5	5
Tarangire National Park, near Englehardt Bridge	1	1
Graphiurus microtis	1	1
Tarangire National Park, near Mabuyu Mingi Campsite	1	1
Graphiurus microtis	1	1
Udzungwa Mountains, Ndundulu Forest.	9	9
Petrodromus tetradactylus	9	9
Udzungwa Mts, Ndundulu Forest, 9 km E Udekwa	13	13
Congosorex phillipsorum	9	9
Crocidura munissii	4	4
Udzungwa Mts, New Dabaga/Ulangambi Forest Reserve	1	1
Suncus megalura	1	1
Udzungwa Mts, West Kilombero Scarp Forest Reserve	19	19
Congosorex phillipsorum	15	15
Suncus megalura	4	4
Udzungwa Mts, West Kilombero Scarp Forest Reserve, Ndundulu Forest	2	2
Beamys hindei	1	1
Petrodromus tetradactylus	1	1
Udzungwa Mts, West Kilombero Scarp Forest Reserve, Nyumbanitu Forest	2	2
Petrodromus tetradactylus	1	1
Rhinolophus deckenii	1	1
Udzungwa Mts, West Kilombero Scarp Forest Reserve, Ukami Forest	2	2

Rhinolophus deckenii			2	2
Uluguru Mts, Uluguru North Forest Reserve, sat camp 2			1	1
Beamys hindei			1	1
Uluguru Mts, Uluguru South Forest Reserve, site 5			1	1
Beamys hindei			1	1
Urban Region, West District, Dole, Masingini Forest Station			9	9
Nycteris thebaica			9	9
Vicinity Chitinji Camp, Zomba Plateau		1		1
Crocidura olivieri		1		1
Total	113	170	427	710

Printed by Books on Demand GmbH, Norderstedt / Germany